W9-CWK-230

5/0
Access

WHOLE NUMBERS AND INTEGERS

GLOBE FEARON EDUCATIONAL PUBLISHER
A Division of Simon & Schuster
Upper Saddle River, New Jersey

Executive Editor: Barbara Levadi
Editors: Bernice Golden, Lynn Kloss, Bob McIlwaine, Kirsten Richert
Production Manager: Penny Gibson
Production Editor: Walt Niedner
Interior Design: The Wheetley Company
Electronic Page Production: Curriculum Concepts
Cover Design: Pat Smythe

Printed in the United States of America   2 3 4 5 6 7 8 9 10 99 98

ISBN 0-8359-1544-1

GLOBE FEARON EDUCATIONAL PUBLISHER
A Division of Simon & Schuster
Upper Saddle River, New Jersey

CONTENTS

TO THE STUDENT

Access to Math is a series of 15 books designed to help you learn new skills and practice these skills in mathematics. You'll learn the steps necessary to solve a range of mathematical problems.

LESSONS HAVE THE FOLLOWING FEATURES:

❖ Lessons are easy to use. Many begin with a sample problem from a real-life experience. After the sample problem is introduced, you are taught step-by-step how to find the answer. Examples show you how to use your skills.

❖ The *Guided Practice* section demonstrates how to solve a problem similar to the sample problem. Answers are given in the first part of the problem to help you find the final answer.

❖ The *Exercises* section gives you the opportunity to practice the skill presented in the lesson.

❖ The *Application* section applies the math skill in a practical or real-life situation. You will learn how to put your knowledge into action by using manipulatives and calculators, and by working problems through with a partner or a group.

Each book ends with *Cumulative Reviews*. These reviews will help you determine if you have learned the skills in the previous lessons. The *Selected Answers* section at the end of each book lists answers to the odd-numbered exercises. Use the answers to check your work.

Working carefully through the exercises in this book will help you understand and appreciate math in your daily life. You'll also gain more confidence in your math skills.

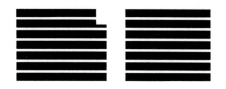

PLACE VALUE

Vocabulary

whole number: a number—such as 0, 1, 2, ...

digit: a symbol used to write a number

place value: the value a digit's position has in a number

According to the 1990 census, the population of the United States was **248,709,873**. This number is read as two hundred forty-eight million, seven hundred nine thousand, eight hundred seventy-three.

The value of any **digit** in a whole number depends on its position, or place, in that number. Each position in the number has its own **place value**.

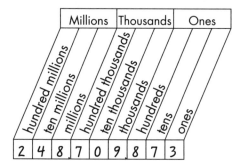

A place-value chart shows the value of each digit. Notice that the digits **8** and **7** each appear twice in the number, but the value of each **8** and **7** is very different.

    2 4 8, 7 0 9, 8 7 3   2 4 8, 7 0 9, 8 7 3

This place is millions.   *This place is hundreds.*
The value is 8 millions,   *The value is 8 hundreds,*
or 8,000,000.   *or 800.*

Notice how commas are used to break the number into groups of three digits.

Guided Practice

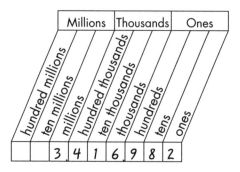

1. What is the value of the digit **3** in the number 3,416,982?

   a. The 3 is in the _____*millions*_____ place.

   b. The value is 3 _____*million*_____ or _____*3000000*_____ .
                           (word)   (number)

2. What is the value of the digit **1** in the number 3,416,982?

    a. The 1 is in the _____ place.

    b. The value is 1 *ten thousand* or _____ .
                        (word)   (number)

Exercises

Write the place of the underlined digit in each number.

3. 60<u>5</u>,386   **4.** <u>2</u>1,312   **5.** 2<u>7</u>0,126,514

_____   _____   _____

6. <u>1</u>22,641   **7.** <u>8</u>1,346,721   **8.** 251,6<u>8</u>2

_____   _____   _____

Write the value of the underlined digit in each number.

9. <u>6</u>,312

_____ or _____

10. <u>8</u>5,562

_____ or _____

11. <u>3</u>,186,342

_____ or _____

12. <u>8</u>7,782,110

_____ or _____

13. 6<u>4</u>2,187

_____ or _____

14. 12,<u>3</u>42,805

_____ or _____

Application

Use a calculator to find each answer. Then turn the calculator upside down. Write the word that appears.

15. Display 8,892.
    Subtract 4 thousands.
    Subtract 9 tens.
    Subtract 3 hundreds.
    Add 4 ones.

    Answer: _____ Word: _____

16. Display 9,517.
    Subtract 2 hundreds.
    Subtract 3 ones.
    Add 2 tens.
    Subtract 8 thousands.

    Answer: _____ Word: _____

17. Display 8,184,216.
    Add 2 hundred-thousands. Add 2 thousands.
    Subtract 1 ten. Subtract 3 millions.
    Add 4 hundreds. Subtract 1 ten-thousand.

    Answer: _____ Word: _____

ROUNDING WHOLE NUMBERS

Vocabulary

rounding: expressing a number to the nearest ten, hundred, thousand, and so on

Reminder

When rounding to a certain place, remember to change all the numbers after that place to zeros.

The actual attendance at the 1994 Major League Baseball All-Star Game was 59,568 people. The newspaper's figure was obtained by **rounding** the number to the nearest ten thousand.

DAILY TRIBUNE

All-Star Attendance Nears 60,000

No matter how large or small the number, the rules for rounding are the same.

To round a number, first find the place being rounded to. Here, the number was to be rounded to the ten-thousands place.

For example, look at the number 59,568.

59, 568 is between 50,000 and 60,000. To figure out which one it is closer to, you look at the digit to the right of the place you are rounding to. If the digit is 5 or greater, round the number **up**. If the digit is less than 5, round the number **down**. Here, 9 is greater than 5, so the number is **rounded up** to 60,000.

$$59,568 \rightarrow 60,000$$

Guided Practice

1. Round 1,452,625 to the nearest hundred.

   a. What is the place that is being rounded to?

   _____

   b. Which digit is to the right of this place?

   _____

   c. Should you round up or round down?

   _____

   d. What is the rounded number? __1,452,600__

Round to the nearest ten.

2. 87

3. 41

4. 165

5. 3,095

_____

_____

_____

_____

Round to the nearest hundred.

6. 228

7. 4,444

8. 32,186

9. 456,153

_____

_____

_____

_____

Round to the nearest thousand.

10. 5,555

11. 1,085

12. 660,814

13. 505,005

_____

_____

_____

_____

Round to the nearest ten thousand.

14. 22,365

15. 156,901

16. 105,633

17. 1,206,380

1,210,000

_____

_____

_____

_____

Round to the nearest hundred thousand.

18. 518,972

600,000

19. 652,131

700,000

20. 2,612,465

2,700,000

21. 12,367,872

12,400,000

_____

_____

_____

_____

Application

22. Put yourself in the place of a newspaper reporter describing how the numbers of navy personnel have changed from 1990 to 1993. You want to make the numbers easy to compare. You also want to keep the data as accurate as possible. Here are the actual data.

1990: 579,417   1991: 570,262   1992: 541,883   1993: 509,950
580,000   540,000   510,000

a. Round each of these numbers so that they are accurate but are easy to compare.

1990: 580,000   1991: 570,000   1992: 540,000   1993: 510,000

b. Describe why you rounded the way that you did.

✏️ _____

_____

_____

ESTIMATING SUMS AND DIFFERENCES

Vocabulary

estimate: an answer that is close to the exact answer

Reminder

To round, look at the digit to the right of the place being rounded to. Then follow the rules for rounding up or down. When you estimate, you use mental math to make the calculation easier.

Vocabulary

The symbol ≈ means "is approximately equal to."

An **estimate** may be made when an exact answer is not needed.

There are several strategies used to estimate a sum or a difference. Estimates can vary, depending on the strategy used. When you estimate, you want to use numbers that are close to the original numbers, but are easier to work with.

Problem 1: Li-Young needs to estimate how many tickets were sold to three concerts last month. She knows that 4,050 tickets were sold for the first concert, 700 for the second one, and 4,380 for the third. Li-Young can use one of the following strategies:

Rounding Strategy	Front-End Strategy

Rounding Strategy:

$$4,050 \longrightarrow 4,000$$
$$0,700 \longrightarrow 1,000$$
$$+\ 4,380 \longrightarrow +\ 4,000$$
$$9,000$$

Front-End Strategy:

$$4,050$$
$$0,700$$
$$+\ 4,380$$
$$8,000$$

Add the front digits, estimate the sum of the rest and adjust the estimate.

$$50 + 700 + 380 \approx 1,000$$
$$8,000 + 1,000 = 9,000$$

Estimate: 9,000   **Estimate: 9,000**

Problem 2: 8,937 tickets were sold last year for the same series. Of those, 4,508 tickets were for the first concert. How many were sold for the remaining concerts?

Rounding Strategy	Compatible Numbers Strategy

Rounding Strategy:

$$8,937 \longrightarrow 9,000$$
$$-\ 4,508 \longrightarrow 5,000$$
$$4,000$$

Compatible Numbers Strategy:

$$8,937 \longrightarrow 9,000$$
$$-\ 4,508 \longrightarrow 4,500$$
$$4,500$$

Use numbers that are easy to work with.

Estimate: 4,000   **Estimate: 4,500**

1. Estimate 6,809 + 4,100.

    a. What strategy did you use? _____

    b. What is the estimate? _____

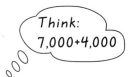

Think:
7,000+4,000

2. Estimate 26,369 – 14,888.

    a. What strategy did you use? _____

    b. What is the estimate? _____

Think:
25-15

Exercises

Estimate. Use rounding or compatible numbers.

3.   356
  602
+ 418

4.   678
+ 80

5.   2,426
  3,850
+ 3,127

6.   280
– 32

7.   2,468
– 1,854

8.   6,741
– 3,112

Estimate. Use front-end estimation.

9.   1,426
  2,306
+ 4,132

10.   846
  120
+ 83

11.   3,612
– 2,380

Choose a strategy to estimate.

12.   3,081
+ 6,112

13.   408
  263
+ 142

14.   6,315
– 1,146

Application

15. What strategy do you prefer to use when estimating? Why?

_____

_____

_____

ADDING WHOLE NUMBERS

Reminder

When renaming, remember:
10 tens = 1 hundred
10 hundreds = 1 thousand
10 thousands =
1 ten thousand

Reminder

When adding, estimate to check the reasonableness of an answer.

Reminder

Before adding several numbers, be sure to correctly align the digits in columns.

Problem 1: The height of the Sears Tower to its antenna is 1,454 feet. Its antenna is 350 feet. What is the height of the entire structure?

Add 1,454 + 350

Add ones.	Add tens. Rename. 10 tens = 1 hundred	Add hundreds.	Add thousands.
1,454 + 350 ___ 4	¹ 1,454 + 350 ___ 04	¹ 1,454 + 350 ___ 804	¹ 1,454 + 350 ___ 1,804

The height of the entire structure is 1,804 feet.
Is the answer reasonable?
Estimate: 1,400 + 400 = 1,800.
The answer is reasonable.

Sometimes an addition requires several renamings.

Problem 2: Add 2,685 + 8,817

Add ones. Rename 12 ones.	Add tens. Rename 10 tens.	Add hundreds. Rename 15 hundreds.	Add thousands. Rename 11 thousands.
¹ 2,685 + 8,817 ___ 2	¹¹ 2,685 + 8,817 ___ 02	¹ ¹¹ 2,685 + 8,817 ___ 502	¹ ¹¹ 2,685 + 8,817 ___ 11,502

2,685 + 8,817 = 11,502

Estimate to check your answer:
3,000 + 9,000 = 12,000
The answer is reasonable.

Problem 3: Add 5,216 + 362 + 2,492 + 86.

Add ones. Rename 16 ones.	Add tens. Rename 25 tens.	Add hundreds. Rename 11 hundreds.	Add thousands.
1 5,216 362 2,492 + 86 6	2 1 5,216 362 2,492 + 86 56	1 2 1 5,216 362 2,492 + 86 156	1 2 1 5,216 362 2,492 + 86 8,156

So, 5,216 + 362 + 2,492 + 86 = 8,156

Estimate: 5,000 + 400 + 2,500 + 100 = 8,000. The answer is reasonable.

Problem 4: Add 23,842 + 91,495.

Add ones. Add tens. Rename 13 tens.	Add hundreds. Rename 13 hundreds. Add thousands.	Add ten-thousands. Rename 11 ten-thousands.
1 23,842 + 91,495 37	1 1 23,842 + 91,495 5,337	1 1 23,842 + 91,495 115,337

So, 23,842 + 91,495 = 115,337

Estimate: 24,000 + 90,000 = 114,000. The answer is reasonable.

Guided Practice

1. Add 6,387 + 18,283.

   a. Which places need renaming?

   _____

   b. Write the sum.

   _____

   c. Estimate to check your answer. _____

   1
   6,387
   + 18,283
   0

2. Add 5,280 + 366 + 1,006 + 433.

    a. Which places need renaming?

        _____

    b. Write the sum.

        _____

$$\begin{array}{r} 5{,}280 \\ 366 \\ 1{,}006 \\ +433 \\ \hline \end{array}$$

    c. Estimate to check your answer. _____

Exercises

Add.

3.
$$\begin{array}{r} 608 \\ +51 \\ \hline \end{array}$$

4.
$$\begin{array}{r} 123 \\ +36 \\ \hline \end{array}$$

5.
$$\begin{array}{r} 315 \\ +62 \\ \hline \end{array}$$

6.
$$\begin{array}{r} 352 \\ +413 \\ \hline \end{array}$$

7.
$$\begin{array}{r} 2{,}632 \\ +122 \\ \hline \end{array}$$

8.
$$\begin{array}{r} 2{,}450 \\ +107 \\ \hline \end{array}$$

9.
$$\begin{array}{r} 3{,}108 \\ +1{,}161 \\ \hline \end{array}$$

10.
$$\begin{array}{r} 4{,}316 \\ +2{,}143 \\ \hline \end{array}$$

11.
$$\begin{array}{r} 3{,}111 \\ 403 \\ +6{,}042 \\ \hline \end{array}$$

12. 20,112 + 16,326 _____

13. 31,181 + 12,506 _____

14. 62,308 + 4,210 _____

15.
$$\begin{array}{r} 182 \\ +56 \\ \hline \end{array}$$

16.
$$\begin{array}{r} 248 \\ +26 \\ \hline \end{array}$$

17.
$$\begin{array}{r} 176 \\ +304 \\ \hline \end{array}$$

18.
$$\begin{array}{r} 273 \\ +146 \\ \hline \end{array}$$

19.
$$\begin{array}{r} 4{,}189 \\ +326 \\ \hline \end{array}$$

20.
$$\begin{array}{r} 3{,}247 \\ +862 \\ \hline \end{array}$$

21.
$$\begin{array}{r} 6{,}324 \\ +1{,}876 \\ \hline \end{array}$$

22.
$$\begin{array}{r} 1{,}324 \\ 5{,}234 \\ 312 \\ +162 \\ \hline \end{array}$$

23.
$$\begin{array}{r} 1{,}611 \\ 850 \\ 38 \\ +2{,}142 \\ \hline \end{array}$$

24. 13,814 + 12,951 _____

25. 31,216 + 5,585 _____

26. 45,812 + 61,968 _____

27. Knowing the basic addition facts is essential when adding mentally. Write the digits 0–9 in the spaces to complete the addends. Use each digit only once. Some digits are written already.

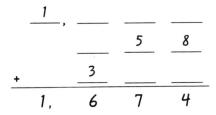

28. A designer needs to purchase carpeting for several floors of an office building. She needs 132 square yards for the first floor, 186 square yards for the second floor, and 288 square yards for the third floor. How many square yards must she purchase?

_____

29. A buyer purchased 3,800 sweaters for the Connecticut store, 1,480 sweaters for the New Jersey store, and 3,820 sweaters for the California store. What is the total number of sweaters purchased?

_____

30. Several people charter a fishing boat. By the end of the day, 497 pounds of mackerel and 1,812 pounds of tuna are caught. How many pounds of fish were caught in all?

SUBTRACTING WHOLE NUMBERS

Reminder

When renaming, remember:
1 hundred = 10 tens
1 thousand = 10 hundreds
1 ten thousand =
10 thousands.

Reminder

When subtracting, estimate to check the reasonableness of an answer.

Last month, a book retailer had 9,486 books in stock. This month's inventory showed that there were 7,823 books in stock. How many books were sold this month?

Subtract 9,486 – 7,823.

Subtract ones. Subtract tens.	Rename thousands to show 10 more hundreds. Subtract hundreds.	Subtract thousands.
9,486 – 7,823 63	8 14 9,486 – 7,823 663	8 14 9,486 – 7,823 1,663

There were 1,663 books sold this month.

Is the answer reasonable?
Estimate: 9,000 – 7,000 = 2,000.
The answer is reasonable.

Sometimes a subtraction requires several renamings.

Problem 1: Subtract 6,234 – 1,847.

Rename tens. Subtract ones.	Rename hundreds. Subtract tens.	Rename thousands. Subtract hundreds.	Subtract thousands.
2 14 6,234 – 1,847 7	12 1 2 14 6,234 – 1,847 87	11 12 5 1 2 14 6,234 – 1,847 387	11 12 5 1 2 14 6,234 – 1,847 4,387

6,234 - 1,847 = 4,387

Estimate: 6,000 - 2,000 = 4,000. The answer is reasonable.

Problem 2: Subtract 1,400 − 265.

Can't subtract ones or tens Rename hundreds.	Rename tens. Subtract ones.	Subtract tens and then hundreds.	Subtract thousands.

      310
1,4̸0̸0
−   265
	9
	3 1̸0 10
	1,4̸0̸0̸
	− 265
	        5
		9
		3 1̸0 10
		1,4̸0̸0̸
		− 265
		    135
			9
			3 1̸0 10
			1,4̸0̸0̸
			− 265
			  1,135

1,400 − 265 = 1,135

Estimate 1,400 − 300 = 1,100. The answer is reasonable.

Problem 3: Subtract 12,440 − 10,505.

Rename tens. Subtract ones.	Subtract tens.	Rename thousands. Subtract hundreds.	Subtract thousands. and then ten thousands.

310
12,4̸4̸0
− 10,505
         5
	310
	12,4̸4̸0̸
	− 10,505
	       35
		1 14 3 10
		1̸2̸,4̸4̸0̸
		− 10,505
		     935
			1 14 3 10
			1̸2̸,4̸4̸0̸
			− 10,505
			  1,935

12,440 − 10,505 = 1,935.

Estimate: 12,500 − 10,500 = 2,000. The answer is reasonable.

Guided Practice

1. Subtract 4,662 − 3,820.

    a. Which places need renaming?

    _____

    b. Write the difference.

    _____

$$\begin{array}{r} 4,662 \\ -\ 3,820 \\ \hline 42 \end{array}$$

2. Subtract 20,355 − 8,648.

    a. Which places need renaming?

    _____

    b. Write the difference.

    _____

$$\begin{array}{r} 20,355 \\ -\ 8,648 \\ \hline \end{array}$$

Exercises

Subtract.

3. 567
    − 42

4. 345
    − 123

5. 674
    − 233

6. 5,366
    − 205

7. 2,809
    − 1,704

8. 4,650
    − 3,420

9. 34,331
    − 10,121

10. 67,381
    − 23,070

11. 16,553
    − 5,422

12. 5,839 − 605 _____

13. 10,676 − 10,424 _____

14. 509
    − 85

15. 826
    − 709

16. 371
    − 170

17. 343
    − 267

18. 740
    − 156

19. 534
    − 480

20. 3,407
    − 316

21. 5,900
    − 1,623

22. 4,345
    − 2,856

23. 23,651
    − 12,805

24. 34,109
    − 30,815

25. 23,700
    − 5,812

26. 356 − 257 _____

27. 6,793 − 805 _____

Application

28. You need to subtract 800 − 693. You don't have a calculator or pencil and paper. Explain how you could solve the problem mentally.

_____

_____

29. Ms. Souza was given an estimate of $26,900 for remodeling her home. The actual cost was $28,817. How much more was the actual cost than the estimate?

_____

30 Dan Marino is one of the all time leading passers in football. Through 1993, his record was 3,219 completed passes out of 5,434 attempted passes. How many of the passes were not completed passes?

31. The height of Mt. Everest in Nepal-Tibet was originally measured as 29,003 feet. But in 1987, an Italian expedition found that its height is 29,108 feet. How much taller is Mt. Everest in the 1987 figure?

MULTIPLICATION FACTS

If there are 3 books on each of 4 tables, how many books are there in all?

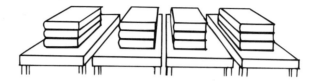

Vocabulary

factors: two or more numbers that are multiplied

product: the result of multiplying

Reminder

Multiply to combine groups that each have the same number.

There are several ways to solve this problem.

Use a counting pattern: 3, 6, 9, 12

Use repeated addition: 3 + 3 + 3 + 3 = 12

Use multiplication, the short form of repeated addition:

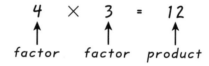

$$4 \times 3 = 12$$

factor   factor   product

4×3 means 4 groups of 3, or 3 + 3 + 3 + 3.

There are 12 books in all.

Guided Practice

1. Tamika grew 5 inches last year. If she continued to grow at this rate, how many inches would she grow in the next 5 years?

   a. Use a counting pattern to solve.

   5, 10, 15, _____

   b. Use repeated addition to solve.

   5 + 5 + 5 + _____

   c. Use multiplication to solve.

   _____

   d. How much will she have grown in 5 years?

   _____

Exercises

Multiply.

2. 9×6

3. 5×4

4. 9×5

5. 8×3

6. 9×1

7. 3×3

8. 6×3

9. 2×9

10. 3×5

11. 9×9

12. 6×6

13. 4×7

14. 2×8

15. 7×7

16. 6×8

17. 3×7

18. $\begin{array}{r} 5 \\ \times\,6 \\ \hline \end{array}$

19. $\begin{array}{r} 6 \\ \times\,2 \\ \hline \end{array}$

20. $\begin{array}{r} 2 \\ \times\,4 \\ \hline \end{array}$

21. $\begin{array}{r} 9 \\ \times\,3 \\ \hline \end{array}$

22. $\begin{array}{r} 4 \\ \times\,4 \\ \hline \end{array}$

23. $\begin{array}{r} 7 \\ \times\,2 \\ \hline \end{array}$

24. $\begin{array}{r} 2 \\ \times\,2 \\ \hline \end{array}$

25. $\begin{array}{r} 4 \\ \times\,9 \\ \hline \end{array}$

Application

26. When hiking, Tyrell travels 4 miles per hour on average. If he hikes for six hours, how many miles has he traveled?

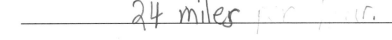
24 miles

27. The Road King bicycle factory hand-assembles 7 bicycles a day. How many wheel assemblies do they use per day?

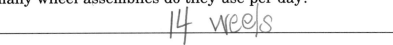
14 weels

28. Ming has triplets. She is making one sweater for each of the triplets. Each sweater will have 6 buttons down the front. How many buttons does Ming have to buy?

18

ESTIMATING PRODUCTS

Reminder

Round the factors to more convenient numbers when you are estimating a product.

Reminder

When estimating, use mental math to make the calculation easier.

Ernesto computed 38×47 on his calculator. The calculator displayed this figure. Is this a likely answer?

When you are using a calculator, sometimes you might press a wrong key. To determine if an answer is likely, or reasonable, use estimation.

To estimate a **product**, change the numbers to ones that are easy to multiply mentally, such as multiples of 10.

$$38 \times 47$$
$$\downarrow$$
$$40 \times 50 = 2,000$$

Since the product of 40×50 is larger than the product of 38×47, the answer Ernesto gets should be some number a little less than 2,000. Ernesto entered something incorrectly.

Remember that an estimate does not necessarily tell you if an answer is correct. It can only tell you if an answer is a reasonable one.

Guided Practice

1. Estimate 8×82.

   a. Change 8×82 to numbers that you can multiply mentally.

   $8 \times$ _____

   b. Estimate 8×82. _____

2. Estimate 53×65.

   a. Change 53×65 to numbers that you can multiply mentally. _____

   b. Estimate 53×65. _____

Estimate each product.

3. 8 × 364

4. 5 × 511

5. 66 × 9

_____

_____

_____

6. 22 × 17

7. 350 × 7

8. 66 × 12

_____

_____

_____

9. 820 × 45

10. 3,678 × 27

11. 44 × 6,217

_____

_____

_____

Application

Estimate to solve.

12. A stadium has 12 sections of seats. Each section has 86 seats. What is the approximate seating capacity of the stadium?

_____

13. A pilot shuttles vacationers from Oahu to Molokai 8 times each week. Each trip is 57 miles. About how many miles does the pilot fly each week between the islands?

_____

14. A migrating humpback whale can swim up to 16 miles an hour. If it could swim at this speed for 24 hours, about how far would it travel in one day?

_____

Use estimation to match the problems in the left hand column with the correct product in the right hand column.

15. 33 × 87 _____ **a.** 20,332

16. 321 × 6 _____ **b.** 2,871

17. 17 × 1,196 _____ **c.** 12,036

18. 14 × 442 _____ **d.** 6,188

19. 3 × 4,012 _____ **e.** 1,926

MULTIPLYING WHOLE NUMBERS

Reminder

A multiple of 10 is the product of any number times 10.

Reminder

When multiplying, estimate to check the reasonableness of an answer.

Reminder

While calculating, be sure to correctly align the digits of each of the partial products.

Three people want to drive to Cleveland from Miami. They plan to share the driving equally over a 24-hour period. If they drive at a constant rate of 55 miles per hour for 24 hours, how far will they travel?

Multiply 24×55.

Multiply by ones.	Multiply by tens.	Add.
4×55 = 220	20×55 = 1,100	

$$\begin{array}{r} 55 \\ \times\ 24 \\ \hline 220 \end{array} \qquad \begin{array}{r} 55 \\ \times\ 24 \\ \hline 220 \\ 1{,}100 \end{array} \qquad \begin{array}{r} 55 \\ \times\ 24 \\ \hline 220 \\ +\ 1{,}100 \\ \hline 1{,}320 \end{array}$$

220 ← *partial product*
+ 1,100 ← *partial product*

They will travel 1,320 miles.

Is the answer reasonable? Estimate: $20 \times 60 = 1,200$. The answer is reasonable.

When multiplying larger numbers, use the same method as for smaller numbers.

Problem 1: Multiply 321×622.

Multiply by ones.	Multiply by tens.	Multiply by hundreds.	Add.
1×622	20×622	300×622	

$$\begin{array}{r} 622 \\ \times 321 \\ \hline 622 \end{array} \quad \begin{array}{r} 622 \\ \times 321 \\ \hline 622 \\ 12{,}440 \end{array} \quad \begin{array}{r} 622 \\ \times 321 \\ \hline 622 \\ 12{,}440 \\ 186{,}600 \end{array} \quad \begin{array}{r} 622 \\ \times 321 \\ \hline 622 \\ 12{,}440 \\ +\ 186{,}600 \\ \hline 199{,}662 \end{array}$$

So, $321 \times 622 = 199,662$

Estimate: $300 \times 600 = 180,000$. The answer is reasonable.

Problem 2: Multiply $318 \times 4,216$.

Multiply by ones. $8 \times 4,216$	Multiply by tens. $10 \times 4,216$	Multiply by hundreds. $300 \times 4,216$	Add.
$\begin{array}{r} 4,216 \\ \times\ \ \ \ 318 \\ \hline 33,728 \end{array}$	$\begin{array}{r} 4,216 \\ \times\ \ \ \ 318 \\ \hline 33,728 \\ 42,160 \end{array}$	$\begin{array}{r} 4,216 \\ \times\ \ \ \ 318 \\ \hline 33,728 \\ 42,160 \\ 1,264,800 \end{array}$	$\begin{array}{r} 4,216 \\ \times\ \ \ \ 318 \\ \hline 33,728 \\ 42,160 \\ +\ 1,264,800 \\ \hline 1,340,688 \end{array}$

$318 \times 4,216 = 1,340,688$

Estimate: $300 \times 4,000 = 1,200,000$. The answer is reasonable.

Problem 3: Multiply $200 \times 3,200$.

There is a shortcut method for multiplying with multiples of 10. Count the total number of zeros and write them in the answer. Then multiply the nonzero numbers.

$\begin{array}{r} 3,200 \\ \times\quad 200 \\ \hline 640,000 \end{array}$ ← *2 zeros here*
← *2 zeros here*
← *Write 4 zeros here.*
Then multiply 2×32.

$200 \times 3,200 = 640,000$

Guided Practice

1. Multiply 422×681.

   a. Multiply by ones.

      $2 \times 681 =$ _____

   b. Multiply by tens.

      $20 \times 681 =$ _____

   c. Multiply by hundreds.

      $400 \times 681 =$ _____

   d. Add, then write the product. _____

$\begin{array}{r} 681 \\ \times\ \ \ 422 \\ \hline 1362 \end{array}$

Multiply.

2. 26
    × 32
    832

3. 51
    × 16
    816

4. 67
    × 21
    916

5. 123
    × 24
    2952

6. 330
    × 44
    14,520

7. 678
    × 77

8. 555
    × 112
    4752

9. 434
    × 600
    269,400

10. 912
    × 367

11. 3,210
    ×   60
    192,600

12. 5,214
    ×   22
    114,708

13. 6,412
    ×   25
    16,300

14. 6,700
    ×   410

15. 5,181
    ×   527

16. 6,340
    ×   190

17. 25 × 80 _____

18. 26 × 121 _____

19. 467 × 411 _____

20. 33 × 1,571 _____

21. 50 × 1,400 _____

22. 222 × 4,521 _____

Application

23. Use a calculator. What number multiplied by itself has a product of:

    a. 3,136   **b.** 3,969   **c.** 8,464

    _____   _____   _____

24. Write about how estimating could be used to find the numbers in exercise 23.

    _____

    _____

25. Virve rides the train twice every day. Each train ride lasts 22 minutes. How many minutes does Virve spend on the train in a year? (Hint: a year has 365 days.)

_____

_____

26. If 4,039,000 children were born one year in the United States, then how many children will be born over 25 years if the rate remains the same?

_____

_____

27. How many children will be born over 48 years if the birthrate remains the same?

_____

_____

28. How many children will be born over 100 years using the same rate?

_____

_____

USING EXPONENTS

You have 2 biological parents. Each of your parents has 2 parents, your grandparents. Each of your grandparents has 2 parents, your great-grandparents. Find each number of biological parents, grandparents, and great-grandparents that you have.

Each number of parents, grandparents, and great-grandparents may found by multiplying. The numbers may also be found in a shortcut method of multiplication which uses **bases** and **exponents**.

Vocabulary

base: a factor that is repeated

exponent: a number that tells how many times another number is used as a factor

Parents: Grandparents: Great-grandparents:

$2 \times 1 = 2$ $2 \times 2 = 4$ $2 \times 2 \times 2 = 8$

$2^1 \leftarrow$ exponent $= 2$ $2^2 = 4$ $2^3 = 8$

$\uparrow$
base

2^1 is read "2 to the first power." 2^1 means that 2 is used as a factor once.

2^2 is read "2 squared," or "2 to the second power." 2^2 means that 2 is used as a factor twice.

2^3 is read "2 cubed," or "2 to the third power." 2^3 means that 2 is used as a factor three times.

The number of parents is 2. The number of grandparents is 4. The number of great-grandparents is 8.

Guided Practice

1. Write the expression $3 \times 3 \times 3 \times 3$ using exponents.

   a. Write the base. _____

   b. Write the exponent. _____

   c. Write $3 \times 3 \times 3 \times 3$ using exponents.

   _____

2. Find the value of 8^2.

   a. Write the numbers that are multiplied.

   $8 \times$ _____

   b. Find the value of 8^2. _____

Exercises

Write each expression using exponents.

3. $5 \times 5 \times 5$ **4.** 4×4 **5.** $6 \times 6 \times 6 \times 6$

_____   _____   _____

Find the value of each expression.

6. 10^2 **7.** 2^5 **8.** 6^3

_____   _____   _____

Application

 To find 23^3 on a calculator, press:

 23   ⊠   23   =   =

Find the value of each expression. Use a calculator.

9. 222^2 **10.** 15^3 **11.** 2^8

_____   _____   _____

12. What is the difference between 2^6 and 6^2? Which is greater? Why?

✎ _____

_____

_____

13. Direct ancestors are parents, grandparents, great-grandparents, and so on. Each level of ancestors is a generation (your parents are one generation and your grandparents are another). Use exponents to figure out the number of ancestors in ten generations.

_____

_____

ESTIMATING QUOTIENTS

quotient: the answer to a division problem

Reminder

Compatible numbers are numbers that are easy to divide mentally.

Estimate a **quotient** when an exact answer is not needed.

Problem 1: A certain auditorium can seat 4,385 people. If there are 73 rows of seats, about how many people can sit in each row?

To estimate 4,385 ÷ 73, use compatible numbers so that you can compute them mentally.

Change 73 to 70. Change 4,385 to 4,200. Now the division can be done mentally.

$$4,200 \div 70 = 60 \longleftarrow quotient$$

About 60 people can sit in each row.

Problem 2: You should also estimate to tell if a **quotient** is likely, or reasonable.

Tawana entered 498 ÷ 6 on the calculator. The display read 83. Is this answer reasonable?

To estimate 498 ÷ 6, change 498 to 480 and divide mentally. 480 ÷ 6 = 80

Since 80 is very close to 83, the answer on the calculator is reasonable.

Guided Practice

1. Estimate 5,407 ÷ 92.

   a. Change 5,407 ÷ 92 to compatible numbers.

   5,400 ÷ _____

   b. Estimate 5,407 ÷ 92. _____

Exercises

Estimate each quotient.

2. $170 \div 8$ _____

3. $324 \div 6$ _____

4. $784 \div 8$ _____

5. $6{,}312 \div 8$ _____

6. $2{,}234 \div 3$ _____

7. $3{,}412 \div 6$ _____

8. $920 \div 18$ _____

9. $642 \div 24$ _____

10. $611 \div 18$ _____

11. $1{,}989 \div 58$ _____

12. $4{,}535 \div 73$ _____

13. $4{,}981 \div 59$ _____

Application

Estimate a reasonable answer for each problem.

14. Ms. Perez is a part-time day care worker. She earned $365 for 5 weeks of work. About how much are her weekly earnings?

_____

_____

15. Kevin works part-time as a dish washer after school. He earned $2,306 in 48 weeks. About how much is his weekly pay?

_____

_____

16. A rural letter carrier earned $865 for an 82-hour pay period. What is the approximate hourly wage?

_____

_____

DIVIDING BY 1-DIGIT NUMBERS

Vocabulary

quotient: the answer to a division problem

remainder: the number left over after dividing

Reminder

Estimate to determine the reasonableness of a quotient.

Reminder

Watch for special cases when you must place a zero in the quotient.

Six college students want to go to Florida during the spring break. The cost for car rental, gas, tolls, and insurance is $462. Assuming they share the travel expenses equally, what is the cost per person?

$$\$467 \quad \div \quad 6 \quad = \quad ?$$
$$\uparrow \qquad\qquad \uparrow \qquad\qquad \uparrow$$

| total cost | number of people | cost per person |

$$\begin{array}{r} 7 \\ 6\overline{)462} \\ -42 \\ \hline 42 \end{array}$$ → *Estimate. How many 6s in 46?*
→ *Multiply, then subtract.*
→ *Bring down the 2.*

$$\begin{array}{r} 77 \\ 6\overline{)462} \\ -42 \\ \hline 42 \\ -42 \\ \hline 0 \end{array}$$
→ *How many 6s in 42?*
→ *Multiply, then subtract.*
→ *The remainder is 0.*

The quotient is 77. Use compatible numbers to check for reasonableness. $480 \div 6 = 80$

Since 80 is close to 77, the quotient 77 is reasonable. The cost per person is $77.

Guided Practice

1. Divide $8{,}109 \div 9$

   a. How many 9s are in 81?

$$\begin{array}{r} 90 \\ 9\overline{)8109} \\ -81 \\ \hline 00 \end{array}$$

   b. Why must a zero be placed in the quotient?

   _____

   c. Complete the division. What is the quotient?

   _____

   d. What is the remainder? _____

2. Divide $3{,}518 \div 7$

    a. What is the quotient? _____

    b. What is the remainder? _____

Exercises

Divide.

3. $2\overline{)75}$ **4.** $5\overline{)28}$ **5.** $6\overline{)126}$

6. $4\overline{)804}$ **7.** $7\overline{)567}$ **8.** $6\overline{)384}$

9. $2\overline{)1{,}212}$ **10.** $3\overline{)3{,}939}$ **11.** $3\overline{)2{,}237}$

12. $827 \div 7$ _____   **13.** $7{,}703 \div 8$ _____

14. $1{,}975 \div 2$ _____

Application

15. There were 342 people attending a basketball game at the school auditorium. If an equal number of people sat in each of the 3 sections of bleachers, how many people sat in each section?

_____

16. The Cortez family drove from Seattle to Washington, D.C., in 5 days. They drove the same distance each day. The driving distance between these two cities is about 2,860 miles. How many miles did the Cortez family drive each day?

_____

17. The Cortez family made the return trip to Seattle in 4 days. If they drove the same distance each day, how many miles did they drive each day?

_____

DIVIDING BY 2-DIGIT NUMBERS

A theater sold 13,467 tickets during 71 performances of a play. What was the average number of tickets sold for each performance?

Divide 13,467 ÷ 71.

```
        1
71)13467     Estimate. How many 70s in 134?
  − 71       Multiply, then subtract.
    636      Bring down the 6.
```

```
       18
71)13467     Estimate. How many 70s in 636?
  − 71       Think: 70 × 9 = 630, but
    636      71 × 9 = 639. So, use 8.
  − 568      Multiply, then subtract.
    687      Bring down the 7.
```

```
      189
71)13467     Estimate. How many 70s in 687?
  − 71       Multiply, then subtract.
    636
  − 568
    687
  − 639
     48      The remainder is 48.
             Because 48 is more than half of 71,
             round the quotient to 190.
```

Is 190 reasonable? Think: 14,000 ÷ 70 = 200. Since 200 is close to 190, the quotient 190 is reasonable.

On the average, 190 tickets were sold at each performance.

1. Divide 807 ÷ 41.

    a. You could estimate that 8 ÷ 4 = 2. Why isn't the estimate of 2 used in the quotient?

$$\begin{array}{r} 1 \\ 41\overline{)807} \\ -41 \\ \hline 397 \end{array}$$

    b. Complete the divison. 807 ÷ 41 = _____

2. Divide 3,562 ÷ 65. _____

    a. Which numbers will you estimate first? _____

    b. Complete the division. 3,562 ÷ 65 = _____

Exercises

Divide.

3. $23\overline{)96}$

4. $31\overline{)39}$

5. $54\overline{)165}$

6. $78\overline{)4,251}$

7. $58\overline{)1,989}$

8. $41\overline{)79,891}$

9. 61 ÷ 23

10. 7,075 ÷ 86

Application

Use a calculator to solve.

 11. Masako has 330 miles to drive. If he drives 55 miles in each hour, how long will it take him?

_____

12. How long would it take Masako to drive the same distance at 60 miles per hour?

_____

13. How much time would Masako save by traveling at 60 miles per hour rather than at 55 miles per hour?

_____

DIVISIBILITY

Vocabulary

divisible: capable of being divided with no remainder

divisibility rule: a shortcut rule used to determine if a number is divisible by another

Reminder

An even number is any number that ends in 0, 2, 4, 6, or 8.

At a company picnic, 126 people decide to form teams to play baseball. Each team must have 9 players. Can they form teams with equal numbers and not leave anybody out?

Think: $126 \div 9 = 14$. They can form 14 teams and nobody will be left out.

Because the number 126 can be divided by 9 without a remainder, 126 is **divisible** by 9.

Without actually dividing, there is a quick way to know if *any* size number is divisible by 9. Use the **divisibility rule** for 9.

The rule states that if the sum of all digits in a number is divisible by 9, then the number is divisible by 9. Since $1 + 2 + 6 = 9$, and 9 is divisible by 9, then 126 is divisible by 9.

Here are the divisibility rules for 2, 3, 4, 5, 6, 8, 9, and 10.

Divisible by:	The rule is:
2	The ones digit is an even number (0,2,4,6,8).
3	The sum of all digits is divisible by 3.
4	The number formed by the last two digits is divisible by 4.
5	The number ends with 0 or 5.
6	The number is divisible by 2 and by 3.
8	The number formed by the last three digits is divisible by 8.
9	The sum of all the digits is divisible by 9.
10	The number ends with 0.

Guided Practice

1. Use the chart to determine if 1,460 is divisible by 10.

   a. What number does 1,460 end with? _____

   b. Is 1,460 divisible by 10? _____

2. Use the chart to determine if 216 is divisible by 6.

   a. 216 is divisible by 2 because the ones digit ends with the even

     number _____.

   b. When is a number divisible by 3? _____

   c. So, is 216 divisible by 6? _____

Exercises

Is the first number divisible by the second? Write yes or no.

3. 610; 10   **4.** 43; 3   **5.** 342; 9

_____   _____   _____

6. 424; 4   **7.** 544; 8   **8.** 1,241; 4

_____   _____   _____

9. 3,386; 6   **10.** 6,312; 2   **11.** 1,035; 5

_____   _____   _____

Application

Use divisibility rules to solve.

12. A caterer is setting up tables for 146 people. Each table must have 6 people. Is it possible to have only tables of 6 people?

_____

13. A group of 132 people want to form a bowling league. Can they form teams of 4 people and not leave anyone out of the league?

_____

14. Presidential elections occur once every 4 years. A census occurs once every 10 years. If a census occurred in 1990 and a presidential election occurred in 1992, then when will the next two presidential elections come in census years?

_____

_____

SOLVING PROBLEMS WITH MORE THAN ONE STEP

Reminder

Look for information that might help to determine the need to add, subtract, multiply, or divide.

Bob needs 6 tomatoes for a salsa recipe. One store sells tomatoes 3 to a package for $1.69. Another sells them for 40¢ each. Which is the better buy and what, if any, is the savings?

Sometimes a problem requires more than one step to solve. You may even need to do some combination of addition, subtraction, multiplication, and division to find the solution.

Begin by making an overall rough plan of what you will do. In this case, you will find the cost of 6 tomatoes at each store, and then you will compare the prices.

First, calculate the cost of 6 tomatoes at each price.

$$2 \times \$1.69 = \$3.38 \ (by \ the \ package)$$

$$6 \times \$0.40 = \$2.40 \ (by \ the \ piece)$$

To determine the better buy, compare the prices for 6 tomatoes.

$$\$3.38 > \$2.40$$

Finally, subtract to determine the savings.

$$\$3.38 - \$2.40 = \$0.98 \ or \ 98¢$$

Buying tomatoes by the piece at the second store is the better buy because you'll save $0.98.

Guided Practice

1. Inez packs tomatoes 3 to a carton. She packs 98 cartons on Monday and 86 cartons on Tuesday. How many tomatoes were packed?

   a. Make a plan. Find the number of cartons packed. Then find the number of tomatoes in the

   _____ .

   b. Find the total number of cartons.

   $$98 + 86 = \text{_____}$$

c. Multiply the number of cartons by 3.

3 × _____ = _____

d. _____ tomatoes were packed.

Exercises

2. Andrew needs to buy 30 grapefruits. One store sells 15 grapefruits for $12, while another store sells a case of 30 grapefruits for $25. Which buy would be better and by how much?

_____

3. Tamika makes blankets. It takes her 5 hours to make each blanket. She makes $65.00 for every bundle of 4 blankets that she makes. How many hours would it take Tamika to make $325.00?

_____

4. In an Iron Woman race, Roxanne swam 2 miles in 40 minutes, ran 6 miles in 60 minutes, and bicycled 30 miles in 90 minutes. Give the order from slowest to fastest of Roxanne's speed swimming, running, and bicycling. [Hint: calculate how many minutes it took her to travel 1 mile by swimming, by running, and by bicycling. Then compare the answers.]

_____

_____

Application

Use a calculator to solve.

5. The dimensions of the Super Market are 98 feet by 122 feet. The store's monthly rent is $35,868. What is the rental cost per square foot? [Hint: multiply the dimensions to find the area of the store in square feet.]

_____

Use mental math to solve.

6. Ten different circuses are gathered for a circus convention. There are 500 people in each circus. Half of the people in each one are clowns. How many clowns are there in all?

_____

THE NUMBER LINE

Vocabulary

positive number: a number greater than zero

negative number: a number less than zero

integer: any positive or negative whole number

number line: a device that pictorially represents the order of integers

One year, the highest temperature in the United States was at Death Valley. The temperature reached 134°F, or 134 degrees above zero. The same year, the lowest temperature in the United States was at Prospect Creek, Alaska. The temperature dropped to ⁻80°F, or 80 degrees below zero.

134, or ⁺134, is a **positive number**, or a number greater than zero. ⁻80 is a **negative number**, or a number less than zero. The symbols ⁺ and ⁻ indicate a number's relation to zero. They should not be confused with addition and subtraction.

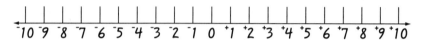

The order of **integers,** that is positive and negative whole numbers, may be shown on a **number line.** On this number line, the numbers to the left of zero are negative. The numbers to the right of zero are positive.

Integers increase in value moving from left to right on this number line.

2 is greater than ⁻4.   *⁻4 is less than ⁻2.*

Write: 2 > ⁻4   *Write: ⁻4 < ⁻2*

Guided Practice

1. Compare ⁻5 and ⁻6. Write < or > .

   a. Which number is farthest to the left on the number line? _____

   b. Which number has the lesser value?

   _____

   c. Write < or >. ⁻5 _____ ⁻6

2. Use the number line to order the integers from least to greatest: ⁻1, ⁻5, ⁻2.

   a. What is the order of these integers on the number line from left to right?

   _____

   b. Write the order of the integers from least to greatest. _____

Exercises

Compare. Write < or >.

3. $^+7$ _____ $^-7$ **4.** $^-4$ _____ $^+1$ **5.** $^+3$ _____ $^+5$

6. $^-5$ _____ $^-4$ **7.** $^-8$ _____ $^+7$ **8.** $^-5$ _____ $^-6$

9. $^+3$ _____ 0 **10.** $^+8$ _____ 0 **11.** 0 _____ $^-10$

Write the integers in order from least to greatest.

12. $^-9, ^-10, ^-6$ **13.** $^+6, 0, ^-5$ **14.** $^+4, ^+2, ^-4$

_____   _____   _____

Application

15. Kiki looked at the thermometer. It read 32 degrees at 9 o'clock. She looked at the thermometer again at 3 o'clock. It read 0 degrees. Did the temperature rise or fall?

_____

16. The temperature reading at noon in Chicago was $^-8°$F. The temperature reading at noon in Duluth was $^-12°$F. Which city had a warmer temperature?

_____

 COOPERATIVE LEARNING **17.** You can compare and order positive and negative numbers without using a number line. Order the following numbers from least to greatest: 4, $^-37$, 84, 227, $^-3$, 473, 821, $^-6$. Make up a rule for comparing positive and negative whole numbers.

_____

_____

_____

_____

_____

ADDING INTEGERS ON THE NUMBER LINE

Jakeela earned $15 babysitting. She owed her mother $4 and her brother $6. What money, if any, will she have after her debts are paid?

Earnings may be shown as positive numbers with or without a + symbol. Money owed, or debts, are shown as negative numbers with a − symbol. You can use addition to find out what she still owes or what she has left after she uses her earnings to pay her debts. Write

$$^+15 + {}^-4 + {}^-6, \text{ or } 15 + {}^-4 + {}^-6.$$

The addition of integers may be shown on a number line. The direction of the arrow depends on the sign of the number. Positive numbers are shown with arrows to the right. Negative numbers are shown with arrows to the left.

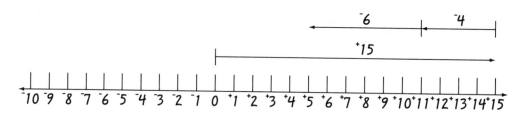

To show 15, draw an arrow from 0 to the right 15 units. To add ⁻4, draw an arrow from 15 to the left 4 units. To add ⁻6, continue the arrow from ⁻4 to the left 6 units. This final number, 5, is the sum.

Jakeela has $5 after settling her debts.

Guided Practice

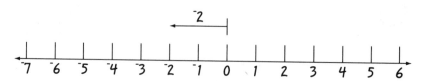

1. Find ⁻2 + 5 on the number line.

  a. Show ⁻2. Start at 0. Draw an arrow 2 places to the left.

  b. Start at ⁻2 and draw an arrow 5 places to the right.

  c. Write the sum. ⁻2 + 5 = _____

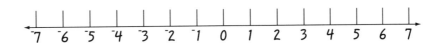

2. Find ⁻1 + ⁻3 + ⁻2 on the number line.

    a. Where did you start and end the arrow to show ⁻1?

    from _____ to _____

    b. Where did you start and end the arrow to show + ⁻3?

    from _____ to _____

    c. Where did you start and end the arrow to show + ⁻2?

    from _____ to _____

    d. Write the sum. ⁻1 + ⁻3 + ⁻2 = _____

Exercises

Find each sum on the number line. Use a pencil so that you can erase and use the number line again.

3. 7 + 2

_____

4. 3 + 2 + ⁻1

_____

5. ⁻2 + 1

_____

6. ⁻6 + 4 + ⁻4

_____

7. 2 + ⁻7 + 5

_____

8. ⁻1 + ⁻4 + ⁻3

_____

Application

COOPERATIVE
LEARNING

9. Play a game with a partner in which one person picks a number between 5 and ⁻5 (including 5 and ⁻5) and the other person tries to create that number by adding at least 2 of the following numbers: 1, ⁻2, 4, and ⁻8.

_____

10. Aside from using a number line, how can you tell whether a sum is positive or negative?

_____

_____

ADDING INTEGERS

Reminder

Be sure to include the negative sign ($^-$) if the sum is a negative integer.

During a football game, a running back ran six plays. He gained 4 yards, lost 3 yards, lost 5 yards, gained 5 yards, lost 6 yards, and gained 2 yards. What was his total loss or gain of yards for the game?

The yardage may be found by adding integers. A gain of yards is shown as a positive integer. A loss of yards is shown as a negative integer.

$$4 + {}^-3 + {}^-5 + 5 + {}^-6 + 2$$

One way to find the sum of several integers is to make groups of two integers and then add:

$$(4 + {}^-3) + ({}^-5 + 5) + ({}^-6 + 2)$$
$$\downarrow \qquad\qquad \downarrow \qquad\qquad \downarrow$$
$$1 \quad + \quad 0 \quad + \quad {}^-4 = {}^-3$$

He had a total loss of 3 yards for the game.

You can also find the sum of several integers by adding all the positive integers first. Then add all the negative integers. Then add them together to find the final sum.

$$(4 + 5 + 2) + ({}^-3 + {}^-5 + {}^-6)$$
$$\downarrow \qquad\qquad\qquad \downarrow$$
$$11 \quad + \quad {}^-14 = {}^-3$$

Guided Practice

1. Find the sum. $^-5 + 4 + {}^-6 + 3$

   a. Make groups of two integers, then add.

   $$({}^-5 + 4) + ({}^-6 + 3)$$
   $$\downarrow \qquad\qquad \downarrow$$
   $${}^-1 \quad + \quad {}^-3 \quad = \underline{\qquad\qquad}$$

   b. Add all positive integers and then all negative integers. Combine sums.

   $$(4 + 3) + ({}^-5 + {}^-6)$$
   $$\downarrow \qquad\qquad \downarrow$$
   $$7 \quad + \quad {}^-11 = \underline{\qquad\qquad}$$

Exercises

Find each sum. Use the method you like best.

2. $6 + {}^-7 + {}^-2 + 1$ **3.** ${}^-2 + {}^-8 + {}^-4 + 14$ **4.** ${}^-8 + 0 + 8 + {}^-2 + {}^-1$

_____   _____   _____

5. ${}^-5 + 7 + {}^-4 + 5 + {}^-3$ **6.** ${}^-4 + {}^-6 + 2 + {}^-5 + 4 + {}^-9$

_____   _____

Application

Write the problem using integers. Then solve.

7. The temperature is 8 degrees in the morning. Throughout the day it rises 3 degrees, then drops 4 degrees, and then drops another 5 degrees by sunset. What is the temperature at sunset?

_____

8. An elevator starts out at the first floor. It goes up 5 floors, descends 2 floors, and then descends another 2 floors. At which floor is the elevator now?

_____

9. Add these numbers: ${}^-29 + 6 + 29 + {}^-3$. What are some different ways to solve this problem? Which way seems easiest and why?

_____

_____

_____

COOPERATIVE **10.** Work with a partner. Write rules for adding positive and negative whole numbers without using a number line.

LEARNING

_____

_____

_____

_____

_____

ADDING MULTIDIGIT INTEGERS

A share of stock was worth $155 on Monday. Over the next three days, the gains and losses were as follows: loss of $12, gain of $18, loss of $10. What was the stock's worth on Thursday?

The stock's worth may be found by adding integers.

$$155 + {}^-12 + 18 + {}^-10$$

You can add larger integers the same way that you added smaller integers.

Make groups of two integers, then add as:

$$(155 + {}^-12) + (18 + {}^-10)$$

$$143 \quad + \quad 8 \quad = 151$$

So, on Thursday, the stock was worth $151.

You can also find the sum of all the positive integers and the sum of all the negative integers. Then the sums can be combined to find the final sum.

$$(155 + 18) + ({}^-12 + {}^-10)$$

$$173 \quad + \quad {}^-22 \quad = 151$$

Guided Practice

1. Find the sum. $85 + {}^-18 + {}^-78 + 12$

   a. Make groups of two integers, then add.

   $$(85 + {}^-18) + ({}^-78 + 12) = \underline{\quad 1 \quad}$$

   b. Add all positive integers and then all negative integers. Combine sums.

   $$85 + 12 = 97, \ {}^-18 + {}^-78 = {}^-96, \ and$$

   $$97 + {}^-96 = \underline{\qquad}$$

   c. Write. $85 + {}^-18 + {}^-78 + 12 = \underline{\qquad}$

Find each sum.

2. $^-23 + ^-30 + ^-12$ **3.** $45 + ^-35 + 27$ **4.** $204 + ^-300 + 200$

_____   _____   _____

5. $^-605 + ^-555 + 125$ **6.** $^-87 + 30 + ^-59 + ^-22$

_____   _____

Application

Write the problem using integers. Then solve.

7. Ms. Cooper bought a stock for $58 in January. In February, the stock lost $11. In March, it gained $20. In April, the stock gained $8. What was the stock's worth in April?

_____

8. Arthur ran four plays in a football game. In the first play, he gained 21 yards. In the second play, he lost 12 yards. In the last two plays, he gained 10 yards and lost 18 yards. What was Arthur's total loss or gain of yards for the game?

_____

9. Determine a good way of adding the following numbers on a calculator: $^-3,850$; 298; $^-960$; $^-5,417$; 76; 33,982. Give the sum. Write about what method you used to add these numbers and why you chose that method.

_____

_____

_____

_____

_____

SUBTRACTING INTEGERS

A chemist recorded a liquid's temperature at ⁻7°C at the beginning of an experiment. Several minutes later the liquid was 5°C. What was the change in temperature?

You can show the temperatures as integers on a number line.

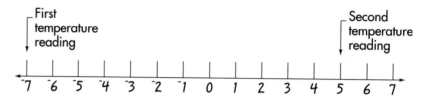

There are 12 units in the positive direction between ⁻7 and 5, so the change in temperature was 12 degrees, or +12 degrees.

You can also find the change in temperature by subtracting the starting temperature reading from the final one:

$$5 - {}^-7 = 12$$

Notice that $5 - {}^-7$ gives you the same result as $5 + 7$. You can rewrite $5 - {}^-7$ as $5 + 7$.

A short-cut rule for subtracting two integers is: add the opposite of the second integer to the first.

Here are some other examples:

$$3 - 6 \qquad {}^-9 - 9 \qquad {}^-8 - {}^-14$$
$$\downarrow \qquad\qquad \downarrow \qquad\qquad \downarrow$$
$$3 + {}^-6 = {}^-3 \qquad {}^-9 + {}^-9 = {}^-18 \qquad {}^-8 + 14 = 6$$

Guided Practice

1. Subtract ⁻9 − 1.

   a. What is the opposite of 1?

   _____⁻1_____

   b. Add the opposite of 1 to ⁻9.

   ⁻9 + ⁻1 = _____

   c. ⁻9 − 1 = _____

Exercises

Subtract.

2. 4 − 5

3. 6 − 8

4. 5 − 7

_____

_____

_____

5. 4 − ⁻2

6. 9 − ⁻8

7. 5 − ⁻9

_____

_____

_____

8. ⁻2 − 6

9. ⁻1 − 5

10. ⁻2 − 7

_____

_____

_____

11. ⁻8 − ⁻8

12. ⁻9 − ⁻2

13. ⁻3 − ⁻4

_____

_____

_____

Application

Use the thermometer like a number line to help you answer the questions below.

14. At midnight the temperature was ⁻10 degrees, at 6 A.M. the temperature was 2 degrees, at noon the temperature was 12 degrees, at 6 P.M. it was ⁻4, and at midnight the next night the temperature was ⁻12 degrees.

a. How much did the temperature change between 6 A.M. and 6 P.M.?

_____

b. How much did it change between midnight and noon?

_____

c. How much did it change between midnight of the first night and 6 P.M.?

_____

SUBTRACTING MULTIDIGIT INTEGERS

Reminder

To subtract two integers, change the second integer to its opposite, then add.

Raivo had a checking account balance of ⁻$23. After making a deposit, his new balance was $45. How much money did Raivo deposit in his account?

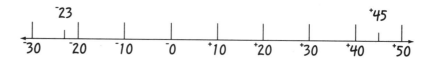

There are two ways that you can think about this problem:

$$^-23 + ? = 45 \text{ or } 45 - {}^-23 = ?$$

When subtracting larger integers, use the same shortcut rule that you learned for subtracting smaller integers.

$$45 - {}^-23 = 45 + {}^+23 = 68$$

↑

Change the second integer to its opposite, then add.

Raivo's deposit was $68.

When subtracting, the shortcut rule always works because addition and subtraction are inversely related. Here are other examples:

$$164 - {}^-13 \qquad {}^-176 - {}^-150 \qquad 44 - 54$$
$$\downarrow \qquad\qquad \downarrow \qquad\qquad \downarrow$$
$$164 + 13 = 177 \quad {}^-176 + 150 = {}^-26 \quad 44 + {}^-54 = {}^-10$$

Guided Practice

1. Subtract $300 - {}^-200$.

   a. What is the opposite of ⁻200? _____⁺200_____

   b. Add the opposite of ⁻200 to 300.

   _____

   c. $300 - {}^-200 =$ _____

Exercises

Subtract.

2. 12 − 22

3. 55 − 35

4. 18 − 22

5. ⁻30 − ⁻22

6. 32 − ⁻29

7. ⁻63 − 15

8. 300 − 400

9. 150 − ⁻200

Application

10. Yuriko thinks that the subtraction of two negative integers should always be a negative integer.

a. Can you think of an example of where this is true? If so, write the example.

b. Can you think of an example where this is not true? If so, then write the example.

c. Do you think Yuriko is right?

11. Cassandra has spilled coffee on her checkbook. She can no longer see what deposits or withdrawals she made. Using the information in the balance column, fill in the missing numbers in the deposit or withdrawal columns.

	Deposits (+)	Withdrawals (−)	Balance	
			$ 322	00
a.			$ 176	00
b.			$ 287	00
c.			$ -25	00
d.			$ 42	00

ADDING AND SUBTRACTING INTEGERS

Reminder

To subtract two integers, change the second integer to its opposite, then add.

Ms. Gates is a math student. She has to calculate the following figures. What is the total?

$$(^-13 + {}^-16) + (29 + {}^-15) - {}^-36$$

When calculating a string of integers that requires addition and subtraction, follow this method:

The first step is to write all the subtractions as addition. The second step is to calculate within the parentheses.

$$(^-13 + {}^-16) + (29 + {}^-15) - {}^-36$$

$$\downarrow$$

$$(^-13 + {}^-16) + (29 + {}^-15) + 36$$

$$\downarrow$$

$$^-29 \quad + \quad 14 \quad + 36 = 21$$

The total is 21.

Guided Practice

1. Find $(12 - 14) - (^-31 + 20)$.

   a. Write all subtraction problems as addition.

   $\underline{\quad (12 + {}^-14) + (31 + {}^-20) \quad}$

   b. Calculate within the parentheses.

   _____

   c. $(12 - 14) - (^-31 + 20) =$ _____

2. Find $28 + (3 - 5) - 12$.

   a. Write all subtraction problems as addition.

   _____

   b. Calculate within the parentheses.

   _____

   c. $28 + (3 - 5) - 12 =$ _____

Find the value of each expression. Change all subtraction problems to addition, then calculate within the parentheses and solve.

3. $^-3 + (8 - 7)$

_____

4. $6 - (^-9 + 5)$

_____

5. $32 - (10 - ^-16)$

_____

6. $(18 - 28) - (^-31 + ^-12)$

_____

7. $^-65 + (37 - ^-41) - ^-12$

_____

8. $28 + (3 + ^-5) + (^-16 - 9)$

_____

9. $(46 + ^-30) - (12 - 14) + 60$

_____

Application

10. Find the missing value that makes the expression true.

$(\underline{\hspace{2cm}} - 25) + (16 - ^-12) = 30$ _____

_____

11. How did you go about finding the missing value? Write about it.

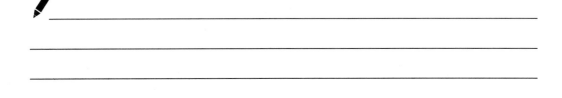

_____

_____

_____

12. Finnegan and Molly play a game of video golf, in which the winner is the player with the lowest score. Finnegan's scores are $^-9$, $^-4$, $^-12$, 24, and $^-18$. Molly's scores are 12, $^-8$, $^-21$, $^-6$, and 22. Who won the game and by how many points?

_____

_____

USING THE CALCULATOR TO ADD AND SUBTRACT INTEGERS

Reminder

Addition and subtraction are inversely related.

Kyela had to calculate a series of debits and credits to find an account's total. What is the account's total?

$$342 + {}^-36 + {}^-723 + 543 + {}^-121$$

Kyela uses a calculator to make the job easier. She knows that on a calculator it is easier to add and subtract all positive integers than it is to add a series of integers with different signs. By changing all of the numbers to positive integers, there will be fewer keys to press. But how can Kyela do this?

Because of the inverse relationship between addition and subtraction, you can change the addition of negative integers to the subtraction of positive integers. So, the series of debits and credits may be rewritten as:

$$342 - 36 - 723 + 543 - 121$$

These numbers may now easily be entered on the calculator as:

$$\boxed{3}\ \boxed{4}\ \boxed{2}\ \boxed{-}\ \boxed{3}\ \boxed{6}\ \boxed{-}\ \boxed{7}\ \boxed{2}\ \boxed{3}\ \boxed{+}$$

$$\boxed{5}\ \boxed{4}\ \boxed{3}\ \boxed{-}\ \boxed{1}\ \boxed{2}\ \boxed{1}\ \boxed{=}$$

The account's total is 5.

Guided Practice

1. Use a calculator to find the value of $16 - 8 + {}^-24 - 53$.

   a. Change the expression to addition and subtraction of all positive integers.

   $$16 - 8 - 24 - 53$$

   _____

   b. Enter: $\boxed{1}\ \boxed{6}\ \boxed{-}\ \boxed{8}\ \boxed{-}\ \boxed{2}\ \boxed{4}\ \boxed{-}\ \boxed{5}\ \boxed{3}\ \boxed{=}$

   c. $16 - 8 + {}^-24 - 53 =$ _____

2. Use a calculator to find the value of
$423 - {}^-120 + {}^-721 - 201$.

    a. Change the expression to addition and subtraction of all positive integers.

    _____

    b. What did you enter on the calculator?

    _____

    c. $423 - {}^-120 + {}^-721 - 201 =$ _____

Exercises

 Use a calculator to find the value of each expression.

3. $56 + {}^-23 - {}^-43$ **4.** $87 - 23 - {}^-12$ **5.** $18 - 23 + {}^-42$

_____   _____   _____

6. $43 - {}^-89 + {}^-67 + 23$ **7.** $23 + 67 - {}^-97 - {}^-16$

_____   _____

8. $450 + {}^-255 - 300 + 111$ **9.** $444 - 711 - {}^-486 + {}^-221$

_____   _____

Application

10. Keeping in mind that addition and subtraction are inversely related, find the value of this expression mentally. (Hint: look at the entire expression before you compute.)

$9 + {}^-6 + 5 + {}^-8 + 6 + {}^-5 + {}^-9 =$

_____

11. Write about how you solved the expression mentally.

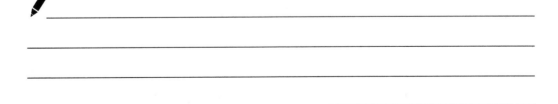

MULTIPLYING INTEGERS

The *Ishmael*, a minisub, descends below sea level at 8 meters per minute. After a dive of 5 minutes, what is the *Ishmael*'s change in depth?

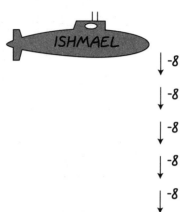

↓ ⁻8

↓ ⁻8

↓ ⁻8

↓ ⁻8

↓ ⁻8

You can find the change in depth by adding or by multiplying with integers:

$$^-8 + {}^-8 + {}^-8 + {}^-8 + {}^-8 = {}^-40$$

or

$$5 \times {}^-8 = {}^-40$$

The *Ishmael*'s change in depth is ⁻40 meters, or 40 meters downward.

You multiply integers just as you do whole positive numbers, but when multiplying integers, it is important to determine whether a product is positive or negative.

Look at the following products.

$$5 \times 8 = 40 \qquad 5 \times {}^-8 = {}^-40$$
$$^-5 \times 8 = {}^-40 \qquad {}^-5 \times {}^-8 = 40$$

The products are the same except for their signs.

The rules for multiplying integers are:

- When both factors have the same sign, the product is positive.
- When the factors have different signs, the product is negative.

Guided Practice

1. Multiply ⁻24 × ⁻8.

   a. What is the product without the sign?

   _____

b. Both factors have the same sign, so the product is ____positive____.

c. Multiply. ⁻24 × ⁻8 = _____

2. Multiply 18 × ⁻20.

   a. What is the product without the sign? _____

   b. The factors have different signs, so the product is _____

   c. Multiply. 18 × ⁻20 = _____

Exercises

Multiply.

3. ⁻4 × 2

4. ⁻5 × ⁻1

5. 6 × 12

6. ⁻13 × 2

7. 20 × ⁻5

8. 12 × ⁻12

9. ⁻35 × 12

10. 22 × ⁻11

Application

Write the problem as both an addition and a multiplication problem using integers, then solve.

11. If the temperature is dropping an average of 4 degrees per week, how much would you expect it to drop over 6 weeks?

   _____

   _____

12. If Jamal loses an average of 3 pounds per week, what will be his projected weight loss for the next 8 weeks?

   _____

   _____

DIVIDING INTEGERS

A deep-sea diver logged a total of ⁻427 feet over the course of 7 dives. What integer represents the average depth per dive?

You can write the problem using integers as:

$$^-427 \div 7 = ?$$

Divide integers just as you do whole numbers, but it is important to remember the rules for determining if a quotient is positive or negative.

The rules for dividing integers are:

- When both integers being divided have the same sign, the quotient is positive.

- When the integers being divided have different signs, the quotient is negative.

Since the integers being divided in the problem ⁻427 ÷ 7 have different signs, the quotient is negative.

$$\text{So, } ^-427 \div 7 = ^-61$$

The average depth per dive was ⁻61 feet.

Guided Practice

Reminder

When dividing integers, remember the rules for determining the sign of the quotient.

1. Divide ⁻360 ÷ ⁻60.

a. What is the quotient without the sign?

_____

b. Both integers being divided have the same sign, so the quotient is positive. ___*positive*___

c. Divide. ⁻360 ÷ ⁻60 = _____

2. Divide 80 ÷ ⁻16.

a. What is the quotient without the sign?

_____

b. The integers being divided have different signs, so the quotient is ___*negative*___

c. Divide. 80 ÷ ⁻16 = _____

3. 24 ÷ 6

_____

4. ⁻30 ÷ 6

_____

5. ⁻54 ÷ ⁻3

_____

6. 123 ÷ ⁻3

_____

7. ⁻225 ÷ 5

_____

8. ⁻242 ÷ ⁻2

_____

9. ⁻248 ÷ 31

_____

10. 108 ÷ 18

_____

11. ⁻256 ÷ 32

_____

Application

Write the problem using integers. Then solve.

12. Mr. Zayas had a loss in the stock market on Tuesday. He had a total loss of $320. He owns 80 shares of the stock. What was the average loss per share of stock?

_____

_____

13. Ms. Kong is trying to figure out the average amount that she withdraws from the bank each month. If her withdrawals for the past 5 months were $60, $72, $45, $23, and $110, what was her average withdrawal?

_____

_____

ORDER OF OPERATIONS

Wei-Chi and Oksana were asked to convert ⁻15°C to degrees Fahrenheit (°F) using the formula °F = °C × 9 ÷ 5 + 32. Which person converted to °F correctly?

Wei-Chi's Way	Oksana's Way
⁻15 × 9 ÷ 5 + 32	⁻15 × 9 ÷ 5 + 32
⁻135 ÷ 5 + 32	⁻135 ÷ 5 + 32
⁻27 + 32	⁻135 ÷ 37
5	⁻3.648

Wei-Chi grouped ⁻15 × 9 and multiplied. Then she divided that figure by 5. Then she added.

Oksana grouped ⁻15 × 9 and multiplied. Then she grouped 5 + 32 and added. Then she divided the figures from her groupings.

They each calculated correctly, but their answers were different. When finding the value of an expression, the order in which you add, subtract, and use other **operations** can make a difference. To avoid the confusion of different answers, people follow a certain order when calculating expressions. The order is called the order of operations. That order is:

- Do all operations in parentheses first.
- Use any exponents from left to right.
- Multiply or divide from left to right.
- Add or subtract from left to right.

Wei-Chi properly followed the order of operations. Her conversion is correct.

Vocabulary

operation: a process such as addition, subtraction, multiplication, or division whereby a mathematical expression is derived from others

Guided Practice

Reminder

When finding the value of an expression with integers, remember to use all the rules involving operations with integers.

1. Find the value of 74 − (14 × 2) + 6.

   a. Do all work in parentheses first.

   14 × 2 = ____28____

   b. Subtract from left to right.

   74 − 28 = ____46____

   c. Add from left to right. 46 + 6 = _____

   d. 74 − (14 × 2) + 6 = _____

56 WHOLE NUMBERS AND INTEGERS

2. Find the value of $(^-18 \times 3) - (16 \div ^-4) \times 2^3$

a. Do all work in parentheses first.

$^-18 \times 3 =$ _____ and $16 \div ^-4 =$ _____

b. Use any exponents.

$2^3 =$ _____

c. Perform multiplication from left to right.

_____ $\times$ _____ $=$ _____

d. Subtract from left to write and write the answer.

_____ $-$ _____ $=$ _____

Exercises

Find the value of each expression. Use the order of operations.

3. $(12 - ^-6) \div 2$

_____

4. $20 \div 4 + 7 - 2$

_____

5. $3 \times 7 + ^-3$

_____

6. $4^2 \times (76 - ^-2)$

_____

7. $(5 + ^-7) \div 2 - 1$

_____

8. $^-6 \times (^-64 \div 2^3)$

_____

9. $(8 - ^-12) \times ^-4 - 12$

_____

10. $3^3 - ^-20 \div 4 \times ^-4$

_____

Application

11. Use your calculator or mental math to complete each expression.

a. $2 \times$ _____ $+ 3^2 = 21$

b. $10^3 \div (9 +$ _____ $) = 100$

12. For problem 11, how did you go about finding the missing values in the expressions? Write about it.

_____

_____

_____

MIXED APPLICATIONS WITH INTEGERS

Reminder

Estimate the answer first. This can help you choose the correct operation.

Poonam takes a bus from 53rd Street to work every morning. The bus stops every eight blocks and she gets off at the fourth stop. If the numbers of the streets go in consecutive order, and the bus moves from the higher to the lower numbered streets, then on what street does Poonam get off the bus?

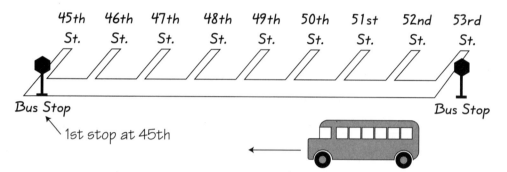

Problems with two or more operations often are solved in two or more steps. But you can often use integers to write these problems in one step. This makes it easier to picture the whole problem at once. The drop in street numbers can be expressed by a negative number. Poonam's bus travels $4 \times {}^-8$ blocks. Where the bus stops can be written as:

$$53 + 4({}^-8)$$
$$\downarrow$$
$$53 + ({}^-32)$$
$$\downarrow$$
$$53 - 32 = 21$$

So, Poonam gets off the bus at 21st Street.

Guided Practice

1. Three boys started a business raking yards. They bought 3 rakes at $7 each, and a tarp for $9. If they earned $60 then how much profit did they make?

   a. Show the expense for the rakes.

   $3 \times$ _____

   b. Add the expense of the tarp.

   $3 \times {}^-7 +$ _____

58 WHOLE NUMBERS AND INTEGERS

c. Add the earnings. $3 \times {}^-7 + {}^-9 +$ _____

d. Find the profit. _____

**Use integers to write the operations for each problem in one step.
Then solve.**

2. Jacques averaged $^-3$ yards a carry for 3 plays in one quarter of a
football game. In the second quarter, he ran one play for 22 yards. What
was his total yardage for the half?

_____

3. Inez had $84 in her checking account. She wrote checks for $18, $23,
and $11. What was her new balance?

_____

4. Missoula, Montana, recorded low temperatures of $^-12°$F, $^-15°$F, $^-13°$F,
and $^-20°$F for four days. What was the average low temperature for the
four days?

_____

5. Izita owes you $11. She pays you $3, borrows $7, and pays you $5. If
she gives you a ten dollar bill are you even?

_____

6. At 5 P.M. the thermometer read $^-10°$F. Over the next 4 hours it dropped
4°F per hour. What was the reading at 9 P.M.?

_____

Application

Use estimation to solve.

7. Aristotle is trying to estimate the depth of a submarine. It starts out
at 197 meters below sea level, then it dives 6 meters per minute for
5 minutes. How far is it below sea level now?

_____

ABSOLUTE VALUE

Vocabulary

absolute value: the distance of a number from 0 on the number line

Lucas wanted to maintain his weight. He didn't want to have too many ups or downs. Lucas kept track month by month of his gains and losses for a year. What was the total change in Lucas's weight over the year?

January	−3	April	−2	July	−3	October	−2
February	+1	May	+4	August	−3	November	−3
March	−4	June	−2	September	+1	December	+2

By looking at the table, you can see that the biggest weight changes took place in March (⁻4) and May (⁺4). The size of a change is called the absolute value of a change. The absolute value is written $|{-4}|$ or $|{+4}|$.

$$size\ of\ change = |{^+4}| = 4\ pounds$$

$$size\ of\ change = |{^-4}| = 4\ pounds$$

You can also picture absolute value on a number line. On the number line, the weight changes can be shown as arrows.

You can see that the size of each change is the distance of that number from zero on the number line. This is the absolute value. For example, ⁻4 is 4 units from zero; ⁺4 is four units from zero.

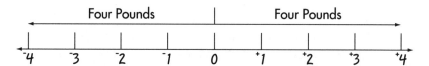

Problem 1. Find the size of change the first three months of the year.

$$|{^-3} + {^+1} + {^-4}|$$
$$|{^-6}| = 6\ lb$$

Problem 2. Find the number of combined pounds gained and lost.

$$|{^-3}| + |{^+1}| + |{^-4}|$$
$$3 + 1 + 4 = 8\ lb$$

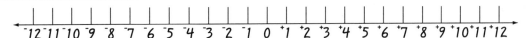

$$\text{-12 -11 -10 -9 -8 -7 -6 -5 -4 -3 -2 -1 0 +1 +2 +3 +4 +5 +6 +7 +8 +9 +10 +11 +12}$$

1. Use the number line above to find: $|{}^+4|$, $|{}^-3|$, $|0|$.

2. Find the value of $|4 - 9|$.

    a. What is the value of $4 - 9$? $4 - 9 =$ _____

    b. What is the absolute value of $^-5$? $|{}^-5| =$ _____

    c. So, $|4 - 9| =$ _____

3. Find the sum of $|{}^+11| + |{}^-5|$

    a. What is the absolute value of $^+11$? $^+11 =$ _____

    b. What is the absolute value of $^-5$? $|{}^-5| =$ _____

    c. What is their sum? $11 + 5 =$ _____

Exercises

Find the value of:

4. $|{}^+4| =$ _____

5. $|{}^-3| =$ _____

6. $|{}^-10| =$ _____

7. $|8 + 2| =$ _____

8. $|9 - 3| =$ _____

9. $|2 - 6| =$ _____

Find the sum or difference.

10. $|{}^+3| + |{}^-5|$ _____

11. $|{}^-2| + |{}^-4|$ _____

Application

12. Yuri is working his way up the face of a mountain. To find handholds, he must sometimes climb down in order to find a new way up. In one afternoon, he climbed up 30 feet, then worked his way down 12 feet, then up 10 feet, then down 28 feet, and finally up 40 feet.

Which number will be greater, the total number of feet Yuri climbed or the number of feet Yuri's destination was from his starting point? How do you know? Show how to find both numbers.

_____

_____

THE COORDINATE PLANE

The grid below is called a coordinate plane. It is divided into four sections, or **quadrants**. Point G is in Quadrant I, and point J is in Quadrant III. Points on either axis are not in any quadrant.

Vocabulary

quadrant: one of the four sections of a coordinate plane that are separated by the axes

x-axis: the horizontal number line

y-axis: the vertical number line

ordered pair: a pair of numbers used to locate a point on a grid

origin: the point where the two lines meet (0,0)

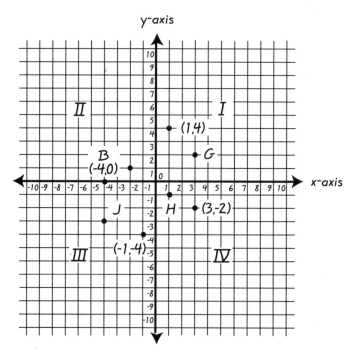

Any point on the grid can be named by using a pair of numbers called an **ordered pair**. To find a point, always start at the **origin,** which is (0,0). The first number tells how far to move to the right or to the left on the **x-axis**. The second number tells how far to move up or down in the same direction as the **y-axis**. How could you use an ordered pair to locate a point on the grid?

To locate the ordered pair (1,4), move one unit to the right and 4 units up.

To find the ordered pair for point B, move 4 units left (⁻4) and move 0 units up or down (⁻4, 0).

Guided Practice

1. Locate point (3,⁻2)

    a. Is (3,⁻2) to the right or left of the origin?

    _____

    b. Is it above or below the x-axis? _____

    c. In which quandrant is point (3,⁻2)? _____

2. What is the ordered pair for point J?

    a. How many units left of the origin did you move? _____

    b. How many units down did you move? _____

    c. Write the ordered pair _____

Exercises

In which quadrant does the point lie?

3. point U _____

4. point S _____

5. point V _____

Locate the following pairs of points and connect them with a line segment.

6. (⁻4,3) and (5,2)

7. (5,2) and (4,⁻2)

8. (4,⁻2) and (⁻6,⁻1)

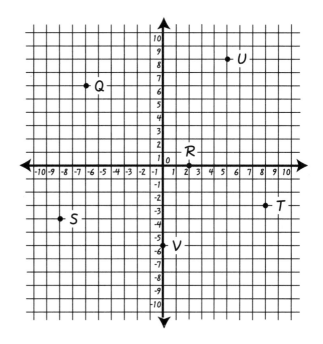

Write an ordered pair for the point.

9. Q _____   **10.** R _____   **11.** T _____

Application

12. Write the ordered pair for the building:

    a. library _____

    b. market _____

13. Name the building at each location:

    a. (2,7) _____

    b. (⁻5,⁻3) _____

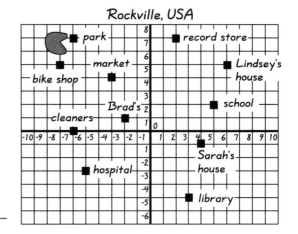

Rockville, USA

14. Lindsey, Sarah, and Brad want to meet at a point that is an equal distance from each of their houses. What are the coordinates for that point? _____

1-3 CUMULATIVE REVIEW

Write the place of the underlined digit in each number.

1. 3<u>6</u>,127   2. 340,<u>1</u>80,105   3. 2<u>4</u>,894,123

_____   _____   _____

Write the value of the underlined digit in each number.

4. <u>1</u>3,567   5. 1,<u>7</u>32,420

_____ or _____   _____ or _____
  (word)   (number)   (word)   (number)

Round to the nearest thousand.

6. 16,840   7. 1,289   8. 312,505

_____   _____   _____

Round to the nearest hundred thousand.

9. 325,615   10. 451,790   11. 21,605,233

_____   _____   _____

On a separate sheet of paper, choose a strategy, then estimate.

12.	1,539	13.	506	14.	6,324	15.	2,343
	− 876		639		− 2,490		2,749
			+ 258				+ 3,542

16. Josef estimated the sum of 1,256 and 820 as 2,000. Amiko estimated the sum of the same two numbers as 1,800. Do you think that both are good estimates? Explain why or why not.

_____

_____

_____

Add.

1. 462
    + 104

2. 328
    + 416

3. 3,413
    + 2,490

4. 3,502
    + 1,869

5. 6,742
    246
    + 4,523

6. 12,468
    + 23,483

7. 35,213 + 809 = _____

8. 603 + 1,231 + 4,075 = _____

Subtract.

9. 432
    − 101

10. 5,386
    − 2,143

11. 4,219
    − 1,068

12. 5,345
    − 4,256

13. 30,567
    − 28,618

14. 32,580
    − 12,881

15. 605 − 83 = _____

16. 5,623 − 1,798 = _____

17. When the Vihmans started out on their trip, the odometer of their car showed 56,209. At the end of the day, the odometer showed 56,571. How many miles did they drive that day?

_____

18. How could you add 97 + 134 mentally? Write about it.

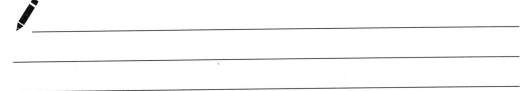

_____

_____

_____

Multiply.

1. $\begin{array}{r} 8 \\ \times 5 \\ \hline \end{array}$ **2.** $\begin{array}{r} 7 \\ \times 6 \\ \hline \end{array}$ **3.** $\begin{array}{r} 9 \\ \times 7 \\ \hline \end{array}$ **4.** $\begin{array}{r} 4 \\ \times 3 \\ \hline \end{array}$

5. $5 \times 4 = $ _____   **6.** $8 \times 8 = $ _____

Estimate each product.

7. 3×85 **8.** 34×62 **9.** 54×67 **10.** 78×49

_____   _____   _____   _____

Multiply.

11. $\begin{array}{r} 24 \\ \times 12 \\ \hline \end{array}$ **12.** $\begin{array}{r} 426 \\ \times 112 \\ \hline \end{array}$ **13.** $\begin{array}{r} 1,912 \\ \times\ \ \ 612 \\ \hline \end{array}$

14. $342 \times 20 = $ _____   **15.** $50 \times 1,120 = $ _____

Write each expression using exponents.

16. $6 \times 6 \times 6 = $ _____   **17.** $18 \times 18 = $ _____   **18.** $4 \times 4 \times 4 = $ _____

Find the value of each expression.

19. $10^3 = $ _____   **20.** $2^4 = $ _____   **21.** $4^6 = $ _____

Solve.

22. If $2^{10} = 1,024$, how could you find 2^{11} mentally? Write about it.

_____

_____

_____

23. Is 5×99 greater or less than 500? How do you know?

_____

_____

LESSONS 10-14 CUMULATIVE REVIEW

Estimate each quotient.

1. $160 \div 4 =$ _____
2. $126 \div 5 =$ _____
3. $2{,}250 \div 7 =$ _____
4. $632 \div 91 =$ _____
5. $924 \div 19 =$ _____
6. $4{,}320 \div 82 =$ _____

Divide.

7. $3\overline{)243}$
8. $6\overline{)156}$
9. $4\overline{)3{,}452}$
10. $34\overline{)612}$
11. $44\overline{)1{,}276}$
12. $52\overline{)19{,}032}$
13. $1{,}470 \div 42 =$ _____
14. $4{,}242 \div 42 =$ _____

Is the first number divisible by the second? Write yes or no.

15. $1{,}242; 9$
16. $3{,}135; 3$
17. $535; 6$

_____ _____ _____

Solve.

18. The Bloom family plans to drive from Albuquerque to Springfield. The driving distances to each of the cities they will travel through are: Albuquerque to Amarillo, 296 miles; Amarillo to Oklahoma City, 260 miles; Oklahoma City to Springfield, 224 miles. How many hours will the trip take if the Blooms average 60 miles per hour?

19. A certain comet is visible from Earth in years that are divisible by 9. Will the comet be in Earth's vicinity in the year 2196?

Compare. Write < or >.

1. 0 _____ ⁻6   **2.** ⁻10 _____ 5   **3.** 6 _____ ⁻6

Write the following integers in order from least to greatest.

4. ⁻6, ⁻3, ⁻9   **5.** ⁻1, 5, ⁻10   **6.** 8, ⁻3, ⁻5

_____   _____   _____

Find each sum on the number line. Use pencil so that you can use the number line again.

7. 7 + ⁻5 = _____   **8.** 7 + ⁻6 + ⁻5 = _____   **9.** 1 + ⁻4 + 3 = _____

Find each sum.

10. 3 + ⁻5 + 2 + ⁻7 = _____   **11.** ⁻8 + ⁻7 + 4 + ⁻3 + 1 = _____

12. 32 + ⁻41 + ⁻18 = _____   **13.** ⁻12 + 51 + ⁻32 + ⁻55 = _____

14. 402 + ⁻123 + 623 = _____   **15.** ⁻177 + 30 + 225 + ⁻87 = _____

Write the problem using positive and negative integers, then solve.

16. Desiree is designing a bridge over a bay. The supports for the bridge will stretch 25 feet below the surface of the water. The total height of the bridge is 67 feet. How many feet does the bridge stretch above the surface of the water?

_____

COOPERATIVE LEARNING

17. How would you explain the difference between 6 and ⁻6 to someone who had no understanding of integers? Write about it.

_____

_____

_____

19-22 CUMULATIVE REVIEW

Subtract.

1. 6 – 7 = _____

2. 4 – ⁻3 = _____

3. 7 – ⁻8 = _____

4. ⁻6 – 8 = _____

5. ⁻23 – 34 = _____

6. 45 – ⁻36 = _____

7. 56 – ⁻32 = _____

8. 326 – 413 = _____

9. ⁻350 – 210 = _____

Find the value of each expression. Rewrite subtraction operations as addition. Do operations inside parentheses first.

10. 4 + (⁻7 – 5)

11. (6 + 32) – 8

12. (⁻12 – 4) + 15

_____

_____

_____

13. 65 + (⁻54 – 13) – 20

14. (23 + ⁻54) – (⁻65 – ⁻23) + 32

_____

_____

 Use a calculator to find the value of each expression.

15. 18 + ⁻132 – ⁻72 = _____

16. 465 – ⁻327 + ⁻422 – 132 = _____

Write the problem using positive and negative integers. Then solve.

17. Ms. Lowenstein had the following debts: electric bill, $67; phone bill, $58; and charge account, $154. She had the following income: paycheck, $636; state tax refund, $96; and federal tax refund, $191. What integer represents the amount she saves (or owes) after settling her accounts?

_____

18. Is the subtraction of one positive integer from another always positive? Explain your answer and support your conclusion with an example.

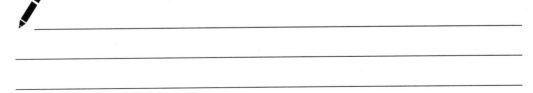

_____

_____

Multiply.

1. $^-5 \times 6 =$ _____

2. $9 \times 7 =$ _____

3. $30 \times {}^-15 =$ _____

4. $^-31 \times {}^-55 =$ _____

5. $18 \times {}^-40 =$ _____

6. $^-25 \times 20 =$ _____

Divide.

7. $32 \div 4 =$ _____

8. $11 \div {}^-11 =$ _____

9. $90 \div 30 =$ _____

10. $^-66 \div {}^-33 =$ _____

11. $^-208 \div 8 =$ _____

12. $^-800 \div {}^-25 =$ _____

Find the value of each expression. Use the order of operations.

13. $^-6 \times 7 + {}^-32 =$ _____

14. $(12 \div {}^-2) - 18 + {}^-21 =$ _____

15. $(5 - 6) \times (32 + 12) =$ _____

16. $(^-22 \div {}^-11) + (^-62 \div {}^-2) =$ _____

Write the problem using integers. Choose the correct operations, then solve.

17. Summers and Wei-Wun played a round of miniature golf. In golf, an eagle is 2 under par (or $^-2$), a birdie is 1 under par (or $^-1$), and bogies are 1 over par (or $^+1$). Par can be represented by 0. During the match, Summers had 4 eagles, 2 birdies, and 4 bogies. Wei-Wun had 3 eagles, 4 birdies, and 3 bogies. They both scored par on the rest of the holes. Who had the lowest score and by how much?

_____

18. In Fargo, North Dakota, the temperature at 5 P.M. was 4°F. Over the next 12 hours the temperature dropped 4°F every hour. Then it rose 3 degrees in one hour. What was the temperature at 6 A.M.?

_____

COOPERATIVE
LEARNING

19. Explain how you found the answer to problems 17 and 18. How do you go about solving problems that involve multiple steps and operations?

_____

_____

LESSONS
27-28 CUMULATIVE REVIEW

Find the absolute value.

1. $|2|$ _____

2. $|7|$ _____

3. $|-324|$ _____

4. $|10 + 6|$

5. $|-17 + 6|$

6. $|-9 - 7|$

_____   _____   _____

Find the sum or the difference.

7. $|+2| + |-8|$

8. $|-17| + |6|$

9. $|82| + |-97|$

_____   _____   _____

10. $|-6| - |4|$

11. $|14| - |-22|$

12. $|-32| - |101|$

_____   _____   _____

Find and mark the following points on the graph.

13. The library is at (2, 7)

14. The bus station is at (-4, 1)

15. The high school is at (5, 4)

16. The city hall is at (0, 0)

17. The park is at (-4, -5)

18. The museum is at (6, -3)

19. The arena is at (3, -6)

20. The theater is at (-6, 3)

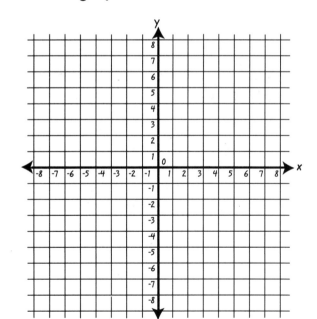

COOPERATIVE LEARNING

21. Draw a square on the graph above. Use (0, 0) as one vertex of the square, then make up coordinates for the three other vertices. Write about the differences and similarities of the coordinates of your square.

_____

_____

_____

ANSWER KEY

LESSON 1 (pages 2–3)
1. **a.** millions **b.** 3 million or 3,000,000
3. thousands
5. ten-millions
7. ten-millions
9. 6 thousands or 6,000
11. 3 millions or 3,000,000
13. 4 ten-thousands or 40,000
15. 4506; gosh
17. 5376606; goggles

LESSON 2 (pages 4–5)
1. **a.** hundred **b.** 2 **c.** down **d.** 1,452,600
3. 40
5. 3,100
7. 4,400
9. 456,200
11. 1,000
13. 505,000
15. 160,000
17. 1,210,000
19. 700,000
21. 12,400,000

LESSON 3 (pages 6–7)
1. **a.** Rounding, Front End, or Compatible Numbers **b.** 11,000 or 10,900
3. R:1,400; C:1,350
5. R:10,000; C:9,500
7. R:0; C:500
9. 7,800
11. 1,300
13. 800
15. Answers will vary.

LESSON 4 (pages 8–11)
1. **a.** ones, tens, thousands **b.** 24,670 **c.** 6,000 + 18,000 = 24,000
3. 659
5. 377
7. 2,754
9. 4,269
11. 9,556
13. 43,687
15. 238
17. 480
19. 4,515
21. 8,200
23. 4,641
25. 36,801
27. 1,047 + 258 + 369
29. 9,100 sweaters

LESSON 5 (pages 12–15)
1. **a.** thousands **b.** 842
3. 525
5. 441
7. 1,105
9. 24,210
11. 11,131
13. 252
15. 117
17. 76
19. 54
21. 4,277
23. 10,846
25. 17,888
27. 5,988
29. $1,917
31. 105 ft

LESSON 6 (pages 16–17)
1. **a.** 20, 25 **b.** 5 + 5 + 5 + 5 + 5=25 **c.** 5 × 5=25
   d. 25 inches
3. 20
5. 24
7. 9
9. 18
11. 81
13. 28
15. 49
17. 21
19. 12
21. 27
23. 14
25. 36
27. 14

LESSON 7 (pages 18–19)
1. **a.** 80 **b.** 640
For exercises 3–14, other estimates are possible.
3. 3,200
5. 630
7. 2,800
9. 40,000
11. 240,000
13. 480 miles
15. b
17. a
19. c

LESSON 8 (pages 20–23)
1. **a.** 1,362 **b.** 13,620 **c.** 272,400 **d.** 287,382
3. 816
5. 2,952
7. 52,206
9. 260,400
11. 192,600

13. 160,300
15. 2,730,387
17. 2,000
19. 191,937
21. 70,000
23. a. 56 **b.** 63 **c.** 92
25. 16,060
27. 193,872,000

LESSON 9 (pages 24–25)
  1. a. 3 **b.** 4 **c.** 3^4
  3. 5^3
  5. 6^4
  7. 32
  9. 49,284
  11. 256
  13. 2^{10}; 1,024 ancestors

LESSON 10 (pages 26–27)
  1. a. 90 **b.** 60
  For exercises 2–18, other estimates are
    possible.
  3. 50
  5. 800
  7. 500
  9. 30
  11. 30
  13. 70
  15. $50

LESSON 11 (pages 28–29)
  1. a. 9 **b.** to represent the zero in the tens
    place **c.** 901 **d.** 0
  3. 37, R1
  5. 21
  7. 81
  9. 606
  11. 745, R2
  13. 962, R7
  15. 114
  17. 715 miles/day

LESSON 12 (pages 30–31)
  1. a. Because 41 won't go into 80 twice
    b. 19, R28
  3. 4, R4
  5. 3, R3
  7. 34, R17
  9. 2, R15
  11. 6 hours
  13. $\frac{1}{2}$ hour

LESSON 13 (pages 32–33)
  1. a. 0 **b.** yes
  3. yes
  5. yes

7. yes
9. no
11. yes
13. yes

LESSON 14 (pages 34–35)
  1. a. cartons **b.** 184 **c.** $3 \times 184 = 552$ **d.** 552
  3. 100 hrs.
  5. $3 per square foot

LESSON 15 (pages 36–37)
  1. a. $^-6$ **b.** $^-6$ **c.** >
  3. >
  5. <
  7. <
  9. >
  11. >
  13. $^-5$, 0, 6
  15. fall
  17. $^-37$, $^-6$, $^-3$, 4, 84, 227, 473, 821 The greater
    the negative number, the less its value; the
    greater the positive number, the greater its
    value.

LESSON 16 (pages 38–39)
  1. c. 3

  3. 9
  5. $^-1$
  7. 0
  9. possible for all numbers between 5 and $^-5$

LESSON 17 (pages 40–41)
  1. a. $^-4$ **b.** $^-4$
  3. 0
  5. 0
  7. 2 degrees
  9. 3
    You can add negative, then positive num-
    bers and combine, or you can notice that $^-23$
    and 23 cancel each other out.

LESSON 18 (pages 42–43)
  1. a. 1 **b.** 1 **c.** 1
  3. 37
  5. $^-1035$
  7. $75
  9. 24, 129 Add all negative numbers or add in
    small groups and combine.

LESSON 19 (pages 44–45)
  1. a. $^-1$ **b.** $^-10$ **c.** $^-10$
  3. $^-2$

5. 6
7. 14
9. ⁻6
11. 0
13. 1

LESSON 20 (pages 46–47)
   1. a. 200 **b.** 500 **c.** 500
   3. 20
   5. ⁻8
   7. ⁻78
   9. 350
   11. a. ⁻$146 **b.** $111 **c.** ⁻$312 **d.** $67

LESSON 21 (pages 48–49)
   1. a. (12 + ⁻14) + (31 + ⁻20) **b.** ⁻2 + 11 **c.** 9
   3. ⁻2
   5. 6
   7. 25
   9. 78
   11. ⁻25 + 28 = 3; 28 + 2 = 30; 27 − 25 = 2

LESSON 22 (pages 50–51)
   1. c. ⁻69
   3. 76
   5. ⁻47
   7. 203
   9. ⁻2
   11. by noticing which positive and negative integers canceled each other out

LESSON 23 (pages 52–53)
   1. a. 192 **b.** positive **c.** 192
   3. ⁻8
   5. 72
   7. ⁻100
   9. ⁻420
   11. ⁻24°

LESSON 24 (pages 54–55)
   1. a. 6 **b.** positive **c.** 6
   3. 4
   5. 18
   7. ⁻45
   9. ⁻8
   11. ⁻8
   13. ⁻$62

LESSON 25 (pages 56–57)
   1. a. 28 **b.** 46 **c.** 52 **d.** 52
   3. 9
   5. 18
   7. ⁻2
   9. ⁻92
   11. a. 6 **b.** 1

LESSON 26 (pages 58–59)
   1. a. ⁻7 **b.** ⁻9 **c.** 60 **d.** $30
   3. $32
   5. Yes, you are even.
   7. 227 meters below sea level

LESSON 27 (pages 60–61)
   1.

   3. a. 11 **b.** 5 **c.** 16
   5. 3
   7. 10
   9. 4
   11. 6

LESSON 28 (pages 62–63)
   1. a. right **b.** below **c.** quadrant IV
   3. I
   5. none
   7.

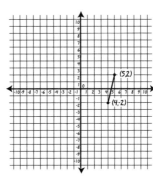

   9. (⁻6, 6)
   11. (8, ⁻3)
   13. a. record store **b.** hospital

CUMULATIVE REVIEW: LESSONS 1–3 (page 64)
   1. thousands
   3. millions
   5. 7 hundred-thousands; 700,000
   7. 1,000
   9. 300,000
   11. 21,600,000
   13. 1,400
   15. 9,000

CUMULATIVE REVIEW: LESSONS 4–5 (page 65)
   1. 566
   3. 5903
   5. 11,511
   7. 36,022
   9. 331
   11. 3,151
   13. 1,949]

15. 522

17. 362 miles

CUMULATIVE REVIEW: LESSONS 6–9 (page 66)

1. 40

3. 63

5. 20

7. 255

9. 3,618

11. 288

13. 1,170,144

15. 56,000

17. 18^2

19. 1,000

21. 4,096

23. Less than, because $5 \times 100 = 500$.

CUMULATIVE REVIEW: LESSONS 10–14 (page 67)

1. 40

3. 300

5. 50

7. 81

9. 863

11. 29

13. 35

15. yes

17. no

19. yes

CUMULATIVE REVIEW: LESSONS 15–18 (page 68)

1. >

3. >

5. ⁻10, ⁻1, 5

7. 2

9. 0

11. ⁻13

13. ⁻48

15. ⁻9

17. The difference between 6 and ⁻6 might be explained as the difference between having 6 objects and owing 6 objects, or giving and taking. It could also be explained visually on a number line.

CUMULATIVE REVIEW: LESSONS 19–22 (page 69)

1. ⁻1

3. 15

5. ⁻57

7. 88

9. ⁻560

11. 30

13. ⁻22

15. ⁻42

17. $644

CUMULATIVE REVIEW: LESSONS 23–26 (page 70)

1. ⁻30

3. ⁻450

5. ⁻720

7. 8

9. 3

11. ⁻26

13. ⁻74

15. ⁻44

17. Wei-Wun, by 1 point less

19. Use order of operations, combine terms before comparing.

CUMULATIVE REVIEW: LESSONS 27–28 (page 71)

1. 2

3. 324

5. 11

7. 10

9. 179

11. ⁻8

13–19.

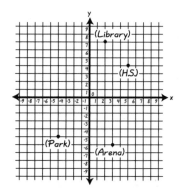

21. The same number will appear combined with 0 in different ways to represent the ordered pair of each vertex.